A dedicated registered nurse in her late 20s with a Bachelor of Science in Nursing (BSN), the author has over two years of experience in various healthcare settings, including hospitals, clinics, and emergency departments. Inspired by her aunt Doris's battle with cancer when the author was 11 years old, she pursued a nursing career to provide compassionate care to patients.

Prior to obtaining her BSN, the author worked as a certified nursing assistant (CNA) for several years, gaining valuable hands-on experience. She also had the opportunity to work in a hospital in Kenya, where she gained exposure to different healthcare systems and cultural perspectives. Since becoming a licensed nurse, she has worked in an observation unit, a clinic, med-surg, and currently in the ER.

With a passion for continuous learning and improvement, the author has collaborated with nurses from diverse backgrounds and levels of experience to create a resource that addresses the challenges and triumphs of the nursing profession.

This book is written in memory of my loving grandparents, Frank Uhde, Karen Uhde, and Jack Sparby, and all of my loved ones and patients that are gone but never forgotten. For all past nurses that are no longer with us, thank you for paving the way.

Mackenzie Sparby

THE NURSING JOURNAL:
WE'RE ONLY HUMAN

AUSTIN MACAULEY PUBLISHERS®

LONDON * CAMBRIDGE * NEW YORK * SHARJAH

Copyright © Mackenzie Sparby 2025

All rights reserved. No part of this publication may be reproduced, distributed, or transmitted in any form or by any means, including photocopying, recording, or other electronic or mechanical methods, without the prior written permission of the publisher, except in the case of brief quotations embodied in critical reviews and certain other non-commercial uses permitted by copyright law. For permission requests, write to the publisher.

Any person who commits any unauthorized act in relation to this publication may be liable to criminal prosecution and civil claims for damages.

The medical information in this book is not advice and should not be treated as such. Do not substitute this information for the medical advice of physicians. The information is general and intended to better inform readers of their health care. Always consult your doctor for your individual needs.

Ordering Information
Quantity sales: Special discounts are available on quantity purchases by corporations, associations, and others. For details, contact the publisher at the address below.

Publisher's Cataloging-in-Publication data
Sparby, Mackenzie
The Nursing Journal: We're Only Human

ISBN 9798891554962 (Paperback)

www.austinmacauley.com/us

First Published 2025
Austin Macauley Publishers LLC
40 Wall Street, 33rd Floor, Suite 3302
New York, NY 10005
USA

mail-usa@austinmacauley.com
+1 (646) 5125767

I would like to first say thank you to my parents, Mark and Dori. They have always been there for me and supported me through all my ups and downs in nursing and throughout my life, and to my family and friends, who are always there for me and always push me to do my best. A special thank you to my grandma, Karen, who, although no longer is here, taught me to always follow my passion, even if others might not agree or believe in the same things I do. Thank you, guys, for everything. Without you, I would not be where I am in life. Also, I would like to say a special thank you to all those who contributed to my book and all the nurses who took time out of their busy schedules to share their stories with me so I could shape my book into something all nurses can use as a tool.

Table of Contents

Introduction	11
Meet the Author	12
Chapter 1: We Are Enough	16
Chapter 2: Burnout	20
Chapter 3: Nursing Ourselves Back to Health	24
Chapter 4: Life Outside of Nursing	30
Chapter 5: Grief/Loss	32
Chapter 6: The Roller Coaster Ride We Call Nursing	36
Chapter 7: Finding the Passion	43
Words of Wisdom	46
References	58

Introduction

To start this book off, I want to start off by defining what life as a nurse means. According to the urban dictionary, living is the pursuit of a lifestyle, and nursing, by definition, is a person trained to care for the sick or infirm. We are in pursuit of a lifestyle as a person trained to take care of the sick or infirm. We are also human, we live a life outside of nursing and we need to take care of ourselves first before we can take care of others, and in order to do that we need to accept that we are all just human and it's okay that we have emotions and it's okay to talk about all the shit we don't like to talk about.

As nurses, we are so used to helping others and listening to other people's problems, that we forget about taking care of ourselves and talking about our problems. I wrote this book for all nurses so that we can understand that we are not alone and it's okay to talk about the things that, for so long, we nurses have pushed down deep inside. Along with learning ways that can help us deal and cope with life again both in our work lives and home lives, in this book, we will be discussing hard topics such as being enough, burnout, nursing ourselves back to health, life outside of nursing, grief/loss, and through all this, finding happiness in our career by sharing the ups and downs that other nurses have experienced and what we can learn from it. I want you as the reader to be able to read and identify with the discussions in this book and help you to understand that your emotions matter, the things you have felt are okay, and most importantly, you are not alone. We stand together.

Use this book as an interactive journal where you can write down your thoughts and feelings that bother you in your life, whether that's at work or at home. Take the time to read and relate to the stories told throughout and allow yourself to learn or relearn how to take care of yourself. Rediscover just how amazing you are; you are not alone in the troubles you face.

"I want you to know that you are not alone in your being alone."

– Stephen Fry.

Meet the Author

If you're going to take the time to read this book, then you should at least know a little something about the person who wrote it. I am 24 years old. I am an RN-BSN and can say, as most nurses reading this probably have, I also have had that one defining moment that made me want to pursue this roller coaster of a career. When I was 11 years old, my aunt Doris got sick with cancer and she was dying. I remember visiting her at the hospital and seeing how caring and loving her nurses were. Even though they were not able to save her, they gave her life with the care they provided, and since that moment on, I have been dedicated to this career. I was a CNA for several years during nursing school and even got the opportunity to work a month in Kenya at their amazing hospital. I got to work closely with nurses from a different culture and learn of the challenges they face. I have now been a licensed nurse for two-and-a-half years. I have worked in many different areas so far in my career, including an observation unit, a clinic, med-surg, and now the ER. I have worked in both bigger hospitals and where I'm now – which is a smaller critical access hospital in my hometown. I can honestly say from my experience, with all different types of nurses from all different areas, at some point in time, we have all struggled with something during our nursing career. I know that I am not some nurse with 50 years of experience, but through my time as a nurse and doing research and reaching out to nurses from all different backgrounds from 1 year of experience to 50 years' experience, I am hoping that with our knowledge combined, this book can be a tool for you, in your future, present or past nursing career.

Why did you become a nurse?

What scares you about nursing?

You are not defined by having knowledge, you are defined by what you do with the knowledge you have.

Chapter 1
We Are Enough

Imposter syndrome – the intense feeling of unworthiness, inadequacy, and experiencing self-doubt. Have you ever had a time when you felt unworthy, inadequate, or not smart enough because you didn't know the answer? Or have been afraid to ask a question because you didn't want it rejected or felt unintelligent? Have you ever felt like an imposter in your field? I have, and that's okay! There is not a single person in this world that knows everything. The same can be said that there is not a single nurse who knows everything there is to do with nursing. We are all unique beings and we all have different knowledge we possess. We were all built with different strengths and weaknesses, but bound by the same passion – the passion of caring for others.

In nursing, we learn most of our skills from being in the field and having hands-on experience, which means constantly having to learn new things which can make us feel like we are never good enough. When in reality, you are good enough whether you are still in nursing school, new to nursing, or an experienced nurse. You're exactly where you're meant to be. We are not meant to know everything. There is a reason we constantly have to renew our education because our field is constantly changing, but there are 20.7 million other nurses learning the same as we are. We all have different strengths and different weaknesses, which is why we can succeed as a team, because together, everyone succeeds more. Don't feel guilty or like an imposter if you don't know something. What we should focus on instead, is our own strengths and weaknesses, our goals, and our accomplishments. By yourself you are strong, but together we are mighty.

A quote by Jay Shetty says, "When we are aware of our strengths and weaknesses, we're more confident, we value ourselves and others' abilities more and we feel less competitive." We need to identify our strengths and weaknesses so we can improve our knowledge. Accepting that we all are going to have different successes and different futures, because success is not defined as just one person, it's all of us working together.

When I was in nursing school, I had a classmate with a photographic memory and for the longest time, I disliked her. I disliked how unintelligent she made me feel. I felt as though I could never be successful if I had to be around people as smart as her. I thought about these things for years, until one day I realized I was judging my success off of her strengths when I should be judging them off of mine. I was paying so much attention to her accomplishments, I had lost focus on the things that I had accomplished. Once I realized that, I was able to look inward and focus on my strengths and weaknesses. Then I could finally understand how important it was to have someone like her in our career. In the end, it's not about who is the most intelligent or who thinks they are the best. We are all there to do one thing, and that's to provide the best care for our patients.

A nurse told me a story once about doing some shadowing on a unit that she was working on; at a bigger hospital, so that way she could better her skills. But when she asked a question, instead of the nurses teaching her, they would belittle her and call her stupid and be angry with her that she did not know something. Their words upset her so much that she actually asked her supervisor not to send her back. Another nurse told me the story of a fellow nurse she worked with who would search for her coworker's personal information online. One day, she came to work and her other coworkers told her that that specific nurse had searched for something personal about her online and sent it to her other coworkers in a group message. She told her supervisor and her supervisor tried to move things around so they would not work together, but that's all that was done. There were no write-ups or warnings, just separation. Now how do we expect our field to change and have the best and brightest nurses when we seek out teaching, friendship, and furthering our education and this type of belittling is constantly occurring? We have to do better to stop harassment and bullying in the workplace.

We have to stop tearing each other down if we ever want to grow as one. We are all on a nursing roller-coaster trying to navigate it the best we can. The least we can do for each other is support and help each other become better at our practice. By encouraging and teaching, not by insulting and belittling. The world can already be such a cruel place. Why tear down the only people who might understand what you're going through? We need to understand that nursing is hard. We are all on the same team and all just working towards the same outcome.

An article by Dr. Renee Thompson, a nurse and bullying expert, discusses how the saying used to be 'nurses eating their young', has now become 'everyone eating everyone'. According to her research, 93% of healthcare employees report witnessing or experiencing bullying behavior. I experienced bullying firsthand from an older nurse when I was brand new in my career. It got so bad that I would leave work every day crying, until one day a friend of mine told me I needed to talk to our supervisor. So I did just that, but there was no change. So, I quit. I physically left a place I loved working because I felt so hopeless.

There will always be bullying no matter what line of work we go into, but that does not mean we have to take it or allow others to go through it. Stand up for yourself and stand up for others, because in the end, we only have each other in this field. Make sure to document when you're being bullied, talk to your supervisors. If they don't listen, take it beyond them, because no one deserves to get bullied. If you are being bullied, you deserve to be heard.

How is it that there is still so much bullying and harassment happening when every year, we have to do training and have policies in place? The answer is simple, because in most cases, people are too afraid to speak up about it, or like in my case, I spoke up about it, and finally did say something and nothing was done about it. As nurses in leadership roles, things such as bullying and harassment need to be taken seriously and proper steps need to be taken to help stop it. As nurses, how about instead of tearing each other down, let's build each other up and help each other realize our true potential. We are successful in our own way and as nurses, we strive to learn more to always further our education and we should never be ashamed or bullied into not succeeding at that. If you have confidence in yourself and your abilities, your strength will follow.

"Always be yourself and have faith in yourself. Do not go out and look for a successful personality and try to duplicate it."

– Bruce Lee.

Strengths

- _____
- _____
- _____
- _____
- _____
- _____
- _____
- _____
- _____

Weaknesses

- _____
- _____
- _____
- _____
- _____
- _____
- _____
- _____

"Weaknesses do not define us, it's what we learn from those weaknesses that do."

Chapter 2
Burnout

I know what you're probably thinking. Why do we always have to talk about burnout? Well, the truth is, the statistics don't lie. According to the American Nurses Foundation, results from a 2020 survey indicated that almost two-thirds of nurses, 62%, experience burnout. It's especially common among younger nurses, with 69% of nurses under 25 reporting burnout. What things can lead to burnout? Longer shifts, shorter orientation, compassionate care, long hours, nursing shortages, nurses eating their young, workplace bullying, not always feeling appreciated for what we have done, struggling with the ethics of patient care, the fact that nursing can be traumatizing and the list could go on.

A fellow colleague of mine experienced burnout firsthand, which actually led him to step away from nursing for 10 years. He told me the story about how he burned out. He was working two jobs: one in an ER and one at a jail. When he worked at the ER, he had to deal with sexual assault patients, and the next day, went and took care of the rapist, and had to provide the same care for both. He finally got to the point where he did not feel it was fair, asking himself why the prisoners deserved to get the same care as the innocent victims. When I asked him if having an outlet or someone to talk to about this struggle would have helped him, he said absolutely, but he was afraid of getting judged not just because of his feelings towards the situation, but because he was a male nurse. As a male nurse, he stated it's even harder to discuss his feelings because he is supposed to be strong and hold his emotions in. This stigma is well-known in the nursing world. I am sure we can all remember male nurses in school being joked about being feminine or soft. Also, the constant stigma that they are men, so why didn't they go on to become doctors, or even in the real world, where they deal with these issues almost every day in the field. As female nurses, we need to start advocating for male nurses and help do away with the stigma. Just because you are male in this field, does not make you feminine or soft. In fact, it does the opposite; it makes you strong. You cared about taking care of people so much and had a true passion for the care only a nurse can provide that you went against the stigmas and all the hateful comments that are made, just to follow your passion. How many people can say that?

Besides the stigma surrounding male nurses, there is also a stigma that surrounds all nurses. We have the constant underlying pressure to keep our emotions in check and are constantly hearing things such as, 'you must see this all the time', and 'you're probably used to it'. So much so that we start to believe it. We become 20-something-year-olds that are not even phased by someone dying in front of our eyes. We learn to just ignore it like it didn't happen, and move on to the next patient. We are so busy taking care of others and being their rock and support that we start to forget early on that it's okay to have feelings and to share things, and we forget that when someone does share these types of things, it's important to listen and not judge.

As nurses, we are taught that the equality of patients is key, even if they are patients who go against everything we stand for. Well, it's okay to be angry, sad, or frustrated with patients, but we must not focus on their beliefs or our beliefs and only focus on the care we provide. I'm not saying to never speak of these feelings again, because we need to reconnect with those feelings and accept our frustration, anger, or sadness and understand that what we felt was okay. It's alright that sometimes it's hard to take care of patients, but what makes you the amazing nurse that you are is, not only that despite your differences, you cared for that patient, you recognized how hard it was and still were able to provide care. Whether that's personally with the patient, or accepting that you might not be the right person for the job, and letting somebody else step in to care for the patient.

The same feeling of frustration, anger, and sadness could be said when Covid 19 hit, where we experienced overwork, staffing shortages, fear about bringing home illness to our family, and to top it all, having to be the only family for those patients because no visitors were allowed. This was one of the most draining times of my career as a nurse, and I'm sure most of you could say the same. It was tiring and almost every day, I wanted to throw in the towel and say that's enough nursing for me. The feeling of being worried about going home every day after work to my family because of the what-ifs. It hurt, I hurt. I was angry. I was so angry: angry about staffing, having to work more hours than I was prepared for, or having to take care of six patients by myself. I was sad not just for our patients who would have to share rooms and listen to other patients pass away, along with not being able to see their families, but I was also sad for their families not being able to see their loved ones during their final moments. I was fearful for myself and my own family, dealing with the constant worry that I would bring sickness home. At first, I used to think, *Mackenzie, you are being selfish. You don't have a right to feel this way. This is your job.* Then I realized, my feelings are what make me the nurse I am. If I didn't feel this way, I wouldn't be human.

It's the human in us that makes us such great nurses and recognizing that it's okay to have these human moments can help us fight the battle of burnout. Understanding that needing to take a step away to focus on our mental and physical well-being is important. We cannot take care of others if we ourselves are not well taken care of. The American Nurses Foundation recommends six things for nurses to help with burnout, including rest, asking for help, getting exercise, eating well, taking a break, and requesting training for coping techniques. Let's be honest, this is easier said than done. So, what else can we do? The first thing is making sure you understand that it's okay not to be okay. It's alright if you need a break. We all need a break at some point in our lives. We also need to understand that sometimes it's okay to be the patient. I'm not talking about being a literal patient. What I mean is, it's time to be the one getting cared for. Understanding that we can talk about the hard things that weigh on us, that's the only way we are going to be able to cope. Our tiredness, sadness, constant second-guessing, and all the whys we face, do not determine our quality of nursing. If anything, it makes us more qualified, because it makes us human.

Compassion fatigue, also known as vicarious or secondary trauma, is the idea that as nurses, we take people's trauma onto ourselves and allow it to become our own. We take care of sick people every day. Some days, we help them. Other days, we can do everything we know to save them and it's still not enough. The feeling of constantly having to take care of others, day in and day out, and not always having the outcome we want, it can overwhelm us, and cause us to get worn out, leading us to being over fatigued. Symptoms of compassion fatigue can consist of feelings of helplessness, reduced feelings of empathy, feeling overwhelmed or exhausted, increased anxiety, withdrawal, and increased conflict in personal relationships. It's important for nurses to understand that they are all going to experience compassion fatigue. Nursing is hard. We battle with emotions every day, and at some point, it becomes physically and mentally draining. We have to create good coping and management strategies to help us process. Practicing activities such as mindfulness or being aware of how you feel about situations throughout the day can help. If you start to feel anxious, focus on controlling your breathing, if you are feeling too overwhelmed, take a moment to yourself to collect your thoughts and make sure to set aside time for yourself and, if needed, reach out to family or friends.

The best thing for us to do, no matter how hard it is, is to talk about the hard things. Whether it's writing it down, talking to other nurses, being in a support group or just saying it out loud, if we don't talk about it, we bottle it up, and we'll eventually collapse. I'm not saying if you do these things you won't ever get burnout, because, at some point in our nursing career, we will all experience burnout. What I am saying is, by talking about the things that are hard and understanding that your thoughts and feelings are okay, we can hopefully push through that feeling of being exhausted, and be able to continue caring for others while caring for ourselves.

Now, as much as talking about issues can help with burnout in a mental aspect, we also need to care for ourselves in a physical and emotional aspect. According to Bradley University, self-care in nurses means taking the necessary steps to reduce stress and take care of our own physical, mental, emotional, and spiritual needs. This is where self-care is important. If we ourselves are not healthy, how can we nurse our patients back to health?

Not only can the emotional side of our job lead to burnout, but the physical aspects can contribute to it as well. As nurses, we work long shifts; most of us working more than three 12-hour shifts a week. Although people outside of this career might think 36 hours is not very much, only we know how draining it truly is. During those 12 hours of having to constantly be on our feet and our nursing brain going nonstop, it can feel like we have just finished running a marathon. I'm sure we have all gone home after working our three shifts and have felt like we could sleep for years. Our career is physically and emotionally demanding. We need to make sure we are taking the right steps to take care of ourselves, not only at work, but at home as well, and allowing ourselves that necessary time to rest and heal.

Use this to write down something that made you smile. That way, when you are struggling, you can look back to this and remember the good times.

    Laugh it out.

Chapter 3
Nursing Ourselves
Back to Health

Did you know that, according to a study done by Davis and colleagues in 2007–2018, nurses were 18% more likely to die from suicide than the general population? I'm sure those numbers have increased since Covid 19. Let that sink in: 18% more likely to commit suicide. This is why self-care is so important in our field. We give so much and take care of so many, we forget to take care of ourselves. Combine that with the toll this job can have on our mental, physical, and emotional health. No wonder that number is so high! We, as healthcare workers, need to understand how draining and overwhelming this job can be and start taking care of ourselves before it's too late.

"It's okay to take time for yourself. We give so much of ourselves to others, we need to be fueled both physically and mentally. If we are in balance, it helps us in all our interactions."

– Faith Hill.

Self-care looks different for everybody. No two people heal and cope the same way. There are many different looks to self-care. Some people paint, listen to music, read, exercise, connect socially, meditate, or journal. But there are also activities such as forming or joining a nurse support group, advocating for debriefings at your workplace, or even just reminding yourself of how awesome you are. I know that last one sounds silly, but it really does help.

For me, during my time in nursing school, grappling with anxiety and depression alongside the demands of my studies, I found solace in a small ritual. I used to draw a black heart on my finger, a quiet reminder that despite my struggles, I was deserving of love—not just from others but from myself. After nursing school, I lost touch with this practice and found myself slipping back into feelings of depression and anxiety. Recently, I've rediscovered the importance of self-affirmation. Before bed each night, I reflect on three things I appreciate about myself or three positive moments from the day. Repeating these affirmations helps me heal and allows my mind to find peace, even if only briefly, amidst the daily chaos.

We all know the feeling of coming home after a long shift, lying in bed, and being unable to shake off the day's events. It's crucial to give our minds a break from the constant demands of nursing and the emotional weight we carry. Jay Shetty, in his book *Think Like a Monk,* suggests that the emotion we go to sleep with often influences how we wake up. He personally reinforces positive affirmations such as feeling relaxed, energized, and focused before bed. While this approach may not resonate with everyone, the key is finding an outlet that supports your self-care and allows you to start each day with a positive mindset.

kind

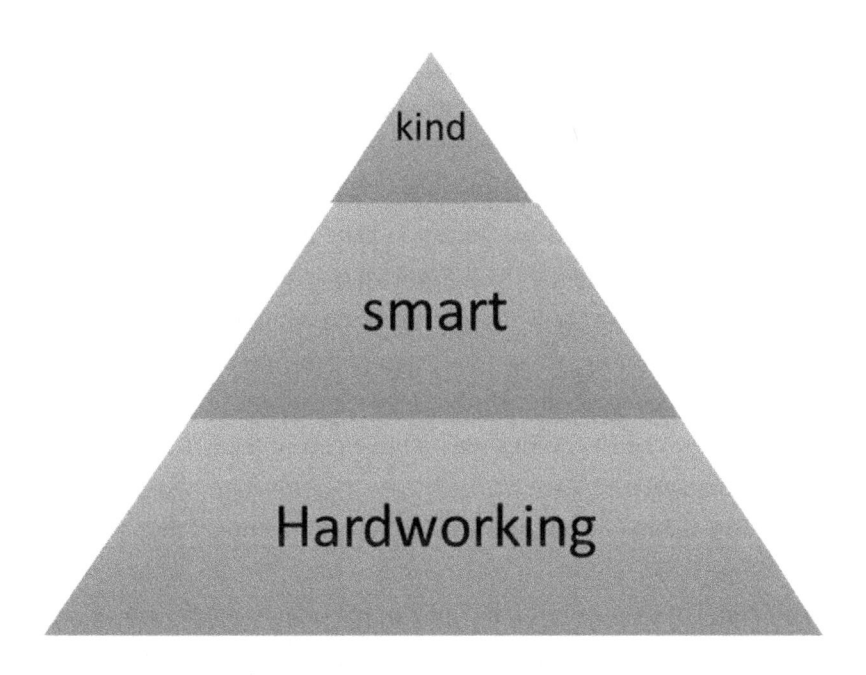

smart

Hardworking

If we don't take care of ourselves, we begin not to be able to care for others. Nurses who don't practice self-care have higher chances of hurting patients and have increased numbers of medication errors, and falls during shifts. Our self-care doesn't just affect us, it affects those that we care for, and it can even affect our families. If we aren't taking care of ourselves, we begin to get run-down and can eventually lose compassion for ourselves and others. This includes both patients and loved ones. We begin to start getting angry at the little stuff or start saying things like, 'nothing bothers me anymore around the workplace', or 'I have run out of compassion for life and the job I do'. This then leads to an increase in depression and anxiety, which then leads to burnout or compassion fatigue. Self-care is so important in your personal life that, when not practiced, it affects your work life. Not only does good self-care support you as a person, but it also supports our patients. If you succeed at taking care of yourself, you will succeed at taking care of others.

A popular saying among monks is, "Make my mind my friend." Our minds are automatically trained to have mental and physical responses to things. We have to learn to manipulate these responses and flip them into something positive. When we are thinking about all the things we could have done better, we should identify all the ways in which we have succeeded. We have to teach our minds to care for ourselves. For example, if you're thinking to yourself, *I could never learn that*, try telling yourself, *I would like to learn that. I will practice as much as I can to learn that.* We control our own minds, which means we can control the outcome of our thoughts.

The times when we are unable to control our thoughts inside our head, we have to let them out. That's where having a good support group comes in. You need to surround yourself with people who each provide something different. That way, the people around you have all different experiences and expertise in all different walks of life, therefore, allowing you to have many different options of people to go to if you're struggling. Rather than asking everyone and getting all different opinions, you might have specific people for different things. Say, you need advice on something but you have a friend that is a good listener but not so good at advice, and one that is good at advice. Then you will choose the one that might actually be able to help you. They are all important in helping with your self-care and your healing journey, but all play a different important role as long as you make sure the people you surround yourself with have a positive impact on your life. The people we surround ourselves with help guide our path to happiness and help with the parts of self-care we can't control ourselves.

If you are struggling with self-care, the first thing you should do is come up with a plan. Look inside yourself to discover your strengths and weaknesses and use this to build goals. The first step is to identify what area of your self-care journey you are struggling with. Is it physically, mentally, spiritually, or personal? Let's say you are very good at making sure to eat right, exercise, and take time to pursue a hobby, but you seem to still be struggling with your mind racing. You're struggling with the mental side of self-care. You found the problem. Now we just need a goal. Instead of playing on your phone for 15 minutes in the morning, you could do 15 minutes of yoga to allow your mind this time to rest. Once you have found where you are struggling, establish small achievable goals that will help you reach your outcome. By completing small obtainable tasks over time, you begin to develop new healthy habits that can help lead to a greater self-help journey.

Nursing is not only a mentally demanding job, but it's also physically demanding. On average, nurses walk four miles in a single shift – that's not including all the other physical labor we do, such as lifting heavy objects, squatting, and bending all day. It's important that once you have tackled the mental side of self-care, you take care of your physical self. I'm not saying that to be a nurse you have to be a supreme athlete, although at times that would probably help! What I am saying is we have to take care of our bodies and form healthy habits. Making sure to eat healthy to keep our nutrition up, stretching regularly to keep our muscles in shape, and doing small things like massages, seeing a chiropractor, and if needed, seeing physical therapist in order to help keep our bodies healthy. Although the mental side of things play a huge part in your success, taking care of your outside self, can help lead to even greater success.

The point of self-care is reflection and creating a balance within yourself that allows a successful home life and a successful work life. If the balance is off, it's up to you to reflect on the steps to your self-care journey and help reset the balance. Self-care is a personal journey, but an important one. As nurses, we always preach to our patients the importance of taking care of themselves and their health. Well, it's about time that we started nursing ourselves back to health.

Iyanla Vanzant once said, "Until you heal the wounds of your past, you will continue to bleed. You can bandage the bleeding with food, with alcohol, with drive, work, with cigarettes, with sex, but eventually it will all ooze through and stain your life. You must find the strength to open wounds, stick your hands inside, pull out the core of the pain that is holding you in your past, the memories, and make peace with them." Sometimes in order to move forward, we have to move backwards. We have to acknowledge the pain, anger, and sadness in order to move past it. We must provide care to ourselves. Things can only overwhelm you if you continue to give them the power to do so, it's time we gain back our control over our own lives, and start healing ourselves from the inside out.

"You are real. You are worthwhile. You are so important both in ways you'll discover, and in ways you'll never see. You send out needed ripples of greatness and kindness in unexpected and accidental ways."

– Jenny Lawson.

learn → to ↓ love ← yourself ← again ↑

If you or someone you know is thinking about suicide, please contact the suicide prevention hotline at 988, or reach out to family, friends, or to your place of work for help.

Things you love about yourself.

- _____
- _____
- _____
- _____
- _____
- _____
- _____
- _____
- _____
- _____

Chapter 4
Life Outside of Nursing

The trick to living a life outside nursing is not asking yourself if it is possible to have both, it's about taking action and finding a way to have both and learning to cope when we struggle too. An old boss of mine told me, the best way to have a successful work and home life is to think of them as two separate things. When she is at work, she does not involve her home life, and when she is at home she leaves work at work. Trust me when I say this is not an easy task. As a nurse, even though we only have to work three to four days a week, we still have things that have to be completed outside of the workplace, such as continuing education, going to courses and seminars, and then there is the fact that we are trained to constantly think of the worst outcomes and how to prepare for them, even when at home, we are always thinking. We have chosen a career with such high physical and mental demands that even when we are home, we either think about work and the things we could have done better, or the things we will have to do when we go back to work.

Sometimes, it feels like there is no win to this never-ending cycle of nursing. Even when we are at home, it feels like there is no rest. So, then what? You have to learn when this constant train of thinking is happening, to find something you like to do to help distract yourself or allow yourself only so much time a day to think about work. Then the rest of the day needs to be spent doing activities you like to do or things that don't involve the nursing brain. This is going to look different for everybody. For example, as a single person living alone, I have a lot more time to distract myself from the hustle of work with things I like to do. Compared to a mom or dad with three kids, even when they aren't at work, they are constantly having to be on the go. Regardless, we all need to take steps to fill our days-off with something satisfying to help take our minds off of all the craziness that goes with our careers.

Setting priorities and boundaries outside of nursing is important. We need to prioritize the important things in our lives so we can organize them into what is most important and what is least important. If family time is at the top of your list and personal time is at the bottom, then family needs to be what you most focus on and what most of your time goes to. Still, allow yourself personal time, but focus on the things that matter most. By doing this, we will have the feeling of success and feel like we are balancing our lives better.

Take time to find joy outside of all the chaos. Find a hobby, or even just take 30 minutes a day to reflect on your thoughts. Understanding that it's okay to say no when work calls, asking you to pick up a shift. Your health is just as important as your patients – mental health is health. Sometimes, even when we try, it can almost be impossible to forget about work. I am sure we can all think of a time when we have all gone home following a tough shift and we get home and our family had no clue what we had to endure at work but expect us to act normal. When in reality, all you want to do is cry, scream, or have a meltdown, but you feel like you can't. You don't want your family to see you like that. Well, in reality, your family would rather see you have a full-blown meltdown than have you push it deep down until one day you explode. Showing these emotions is not going to make your family think any less of you. It will actually do the opposite. It shows just how much you care.

Allow your family to be a part of your pain so that you can move past it. Sometimes, sharing is what allows you to be present in the moment. If something occurs during the day and you go home and all you can think about is that specific thing, how do you get to be present with the then-and-now? Share your feelings, allow yourself to grieve or be angry. That way, you can be present in the things around you.

Besides developing a healthy relationship with your family, it's also important to develop good relationships with your friends outside of work. Make sure to make time to go out and have fun and create a connection. That way, you have a good support system outside of work and home, which will help lead to positive relationships, better health, and help with stress management.

A healthy relationship with family and friends can look different for everybody, but one of the most important things it should include is boundaries. Every day in our career, we get pushed outside of our comfort zones and outside of our boundaries in a way, losing control sometimes of the situation, and the feeling of losing control of our life. By establishing good boundaries at home, such as things that you are willing to talk about or not, or even things like when and if you want to be touched, we gain a little bit of that control we lose. Allowing us to feel more balanced in our everyday lives and making us feel as though the relationships we have or are creating are being formed in a healthy way.

As nurses, we don't always receive the thanks, love, or appreciation at work we deserve. So, we need to focus on the people who show us their support, like our family and friends. You may get angry that your patient did not say thank you, but your parents called that day to check-in. Rather than thinking about the negative of not getting a thank you, appreciate the support and love from your parents. When we learn to appreciate the people around us who already appreciate us, and worry less about what the patients we care for do, it doesn't hurt as badly when we don't get the response from others we were hoping for. Surround yourself with people that care and all the big things that happen during work, and life won't seem so big anymore.

Balancing home life and work life can be a juggling act sometimes, but sometimes all you need is to give yourself an extra hand and allow yourself to lean on others. Remembering to do small things to help with your physical and mental well-being and understanding that it's okay to put yourself first, will help create a balance between your everyday life and your work life, therefore leading to better everyday success.

Chapter 5
Grief/Loss

No one, not even nurses, like to talk about death, or dealing with grief. Why is that? Well, the answer is simple. It sucks! But it doesn't have to. As a nurse, we see death quite often; whether it's expected or unexpected death. We have to deal with the emotional toll of patients taking their last breath in front of our eyes, the last pulse check, knowing there is no good outcome, the last squeeze of the Ambu bag, the moment compressions stop, and the dreaded call of time of death. Then in the aftermath of a death, the family's screams, sobs, and everything you have learned is forgotten in that specific moment, and the reality of death sinks in. So, it's important for us to be able to speak about how it does or doesn't make us feel, and remember that we have no control over death. Death is the only thing guaranteed in this life, but what we can control is the comfort we provide for the patients, families, and ourselves. Whether that's during a debriefing, at work, or with family or friends. Understanding that as a nurse we can seek comfort in the dying process, by understanding that for some patients we are giving them a chance to survive and when we are unsuccessful, we give them the respect and sense of care they deserve. For others, we are giving them the opportunity to die with dignity and not have to suffer anymore.

The thought of death can be viewed in many different ways and we all process death differently. It's okay for you to be super emotional after the loss of a patient and needing to speak about it. It's okay if a patient's death causes a trigger for you, and if you need to step back from that specific thing. If every time you go to put an IO in a patient, it causes you to have a flashback to something traumatic, don't be afraid to ask to switch with someone else. Just because you need a break, doesn't make you any less of a nurse. We're only human, and there is sometimes only so much we can take. If you're the opposite and death doesn't bother you at all, that's okay too. This doesn't mean you are soulless or that you don't care. It just means we are all human and process death in different ways.

A friend of mine, who is also a nurse, and I were having dinner one night and got on the topic of death. We started talking about how we don't remember the faces of every one of our patients that died, and we started questioning if that made us bad nurses. The answer is no. It's okay if you can't remember all the faces, and it's okay if you can. For some, remembering all the faces may help; for others, remembering all the faces will cause pain or triggers. There is no easy way to talk about death or the grieving process, but understanding that no matter how or in what way you deal with the losses, it does not change how good of a nurse or human you are.

Sometimes the hardest thing is not death itself, it can be the family, the age of the patient, or the circumstances regarding death. Just because you don't react to one death, does not mean you won't react to another. It does not mean you cared for one patient more than the other. Losing a 5-year-old patient and losing a 75-year-old patient are going to have different responses, and come with different emotions. Death can happen at any stage of life. We're all just on borrowed time until we run out. Getting to be there for that person, whether that's doing everything we can to save them, or doing everything we can to make death as painless as possible no matter what age or stage of life they are in, is what counts. Emotion comes with the job title, but what we do with that emotion is what matters.

In our line of work, death can be one of two ways: unexpected or expected. Even an expected death, where the family and we as nurses have had time to prepare, can still be just as hard as if the death was unexpected. As nurses, it is our job to support the family through this difficult time, but we also need to support ourselves. For some nurses, practicing empathetic care rather than sympathetic care allows you to still care for the patients through the dying process, but helps keep an emotional block between the patient and the nurse. For others, practicing sympathetic care helps them process what has occurred. Regardless, whatever way you choose to process death and whatever way you grieve, it is okay.

It's natural to form attachments to patients because we care for them daily, addressing all their needs, and supporting them through life's transitions, including death. Many of us can recall moments when a dying patient began experiencing visions—seeing a child in the room or reunited with lost family members—signaling their impending passing. The journey toward death can sometimes involve significant suffering or trauma, which challenges our ability to empathize and requires deep sympathy.

Showing our sympathetic side or using empathy is not a sign of weakness. It shows a sense of vulnerability that is needed in nursing. It allows nurses to relate to their patients and families and helps create a sense of connection that allows better communication, better coping skills, and better patient-focused care. With a sense of connection to your patients, you help create a bond of trust, which allows a better patient care atmosphere. It can allow a healthy connection and happiness for the patient. With a better atmosphere, it can make the process of death easier. It allows the patient to feel as though they were properly cared for, and it allows the nurse to understand and accept that they did everything they possibly could and go through the stages of grief more easily.

In grief, there are six stages and, as a nurse, we are no exception to these stages. We may process and go through the stages quicker, but we still go through them. Stage 1 is denial. In nursing, this looks more like ignoring death and moving on to something else. Stage 2 is anger. Questioning what we could have done differently, and thinking we could have done more. Stage 3 is bargaining. This stage might not necessarily involve the specific patient we lost, but could involve the rule of threes. As we say in nursing, we already lost three patients today or this week, let us not lose anymore. Stage 4 is depression. This would be us finally getting the chance to feel sorry for those we have lost. Stage 5 is acceptance. Finally understanding that we did everything possible for the patient and there was nothing else we could have done. Stage 6 is moving on. We have learned from this experience and are now able to share what we have gone through and the lessons we have learned.

Sometimes, getting through these stages can be difficult. I'm sure we can all remember a time when we've lost a patient but still had to care for five others. Just as we finish tending to the patient who passed away, four call lights suddenly go off: one patient needs assistance to the bathroom, another requests water, a third requires repositioning in bed, and the fourth is in an unexpected situation, perhaps lying on the floor or attempting to climb on the ceiling.

As nurses, when we are juggling so many different things at once, we sometimes don't have time to work ourselves through the stages right away. Sometimes it will take time. I know it seems like everyone expects you to be okay right after these things happen, but you don't have to. It's okay if it takes a while to move along through the steps, and it's alright if it doesn't.

A nurse told me the story of a time when she lost a young patient. It was a very traumatic death. She struggled with it, not just because of what happened, but because the patient was the same age as one of her kids. This caused her to think about what if that was her kid. It became personal. So much so that she kept her own kids home from school the next day because she needed that extra time. When death becomes personal, it's going to take longer to grieve, and if you're reading this thinking you will never let a death at work get personal, you are wrong. At some point in your career, there will be one, and it will take time, but we need that time to process and work ourselves through the grief. In this nurse's case, she just needed that extra day with her own children to help her move past the grief she was feeling. For others, it could take weeks, months or years. The point is that sometimes we get attached and it takes time for us to heal. Other times, it might feel as though we went through the six steps in a matter of seconds. Either way, it's okay. No matter if it takes seconds, weeks, or years, the grief you felt or are still feeling does not define your nursing abilities. We must feel every bit of emotion, no matter how long or little it takes, in order to move past it.

If you are struggling to get through the grieving process, advocating for debriefings at your hospital can make a huge difference in reaching stage six. It allows you to not only talk about the things that went well and the things that didn't, but it also allows us to openly talk about our feelings with people who might be feeling the same thing. Sometimes the best way to grieve is with others. If you are not the type to share your feelings, just support those who are. In some cases, listening can be just as beneficial as talking.

If debriefings aren't enough, don't be afraid to seek help elsewhere, through therapists or support groups that specialize in things like this. Even just reaching out to fellow coworkers to talk about how you're feeling can help. Sometimes, just hearing others share stories of similar situations helps. It's okay to need to lean on someone during difficult times. Sometimes we are only as strong as those around us.

"Nurses are there when the last breath is taken and nurses are there when their first breath is taken. Although it is more enjoyable to celebrate birth, it is just as important to comfort in death."

– Christine Bell.

Chapter 6
The Roller Coaster Ride
We Call Nursing

To say nursing is all about saving lives and praise is not fully true. Although it entails this sometimes, there is also the challenge of patient-nurse ratios, nursing abuse, the difficulty of not feeling appreciated by not only the patients but by their families too, and a constant battle with self-care. For those who have been nurses for a while, I am sure you can relate, but to those new nurses, I would say expect the unexpected, both in a good way and a bad way.

Since Covid 19 hit, till now, nurses have been experiencing a higher nurse-to-patient ratio. Due to short staffing, we have been stuck taking upwards of even six patients to one nurse, and in some cases, even more. Nurses have spoken out about this, and it has even led to nurses going on strike. Not only can this be unsafe for us as nurses, but it is unsafe for the patients. It can be hard enough for one nurse to take care of four patients and then they add two or more to the list. This is a challenge. There are no ifs, and, or buts about it. All we can do is advocate for ourselves and our patients to the best of our abilities and not be afraid to say no.

A nurse told me that Covid was her last straw for being a staff nurse; the feeling of being underappreciated, as though the administration does not care about nurses. She felt as though they didn't care what we were having to go through or deal with, and they did not care about our patient-load or the stress we were facing. While they got to work from home, we had to be on the front lines. They sat back and told us we needed to be better at charting or increased our workload. We were already drowning. Then they give us pizza parties and free meal tickets and act as though this is a good reward. Not saying we are the most important, but we are all part of the machine that makes the hospital go around, and it would be nice if sometimes nurses could feel more appreciated at work. It feels as though they want us to put work above everything else, including family, health, and mental health. And if we don't, it's like there is an unspoken level of guilt applied. The administration then wonders why so many staff nurses are leaving to go traveling. Well, they make more money, and if they don't like something, they just don't have to continue working there once their contract is up. Plus, they can make their own schedules and be present more with family and friends.

Maybe if hospitals focused on us as people and not just as something to help produce more money, there would be higher nurse satisfaction, and they would retain more nurses. Now I am not saying all hospitals are like this or that the administration are awful people. What I am saying is that sometimes we have to put the business part aside and care for the employees as if they actually matter, and can't just be replaced in seconds. Those in administration roles or leadership roles must remember that appreciation can be as simple as a thank you or a job well done as long as it is meaningful, or even advocating for your hospital to give out the DAISY award. As for nurses, the same can be said for your peers. Making sure we show each other we appreciate one another and are thankful for the help we can provide. Appreciation is not just one person's job, we all must learn to appreciate each other in whatever role we are in, and understand that we all play an important part in the system. If our administration is not going to show their appreciation for our work, then it's up to us as nurses to show appreciation for one another. We must create the atmosphere of the value of nurse appreciation, and help lead hospitals down the right path to meaningful recognition of nurses.

There's also the issue of increasing nursing abuse. Nowadays, many hospitals require us to attend self-defense classes to learn how to escape violent situations involving patients. Nursing school doesn't prepare us for this, but incidents of patient violence against nurses are on the rise. In my career, I've been hit, kicked, spit on, and even had my finger dislocated—just some of the physical challenges. The verbal abuse is equally distressing, with daily occurrences of profanity, hurtful remarks, sexual innuendos, racism, and other offensive language. When did this become acceptable? We're often told to brush it off or develop thicker skin, but why should we tolerate this behavior? During my first year and a half as a nurse, I tried to ignore it, but working alongside veteran nurses with over a decade of experience shed light on these issues. They taught me that you have to stand up for yourself or they are going to continue to do these things and believe this type of treatment towards nurses is acceptable. Now I am not saying to yell at them or swear back at them. What I mean is, don't be afraid to tell a patient that what they're saying is inappropriate or they cannot talk to you like that or touch you like that, and if they refuse to stop, it's okay to allow another nurse to take over care. You always have the right to stand up for yourself. Never forget that.

How about sometimes not feeling appreciated for the things we have done? I know most of us are not in this career for appreciation. We do this because we love to help people. That doesn't mean we can't be upset every now and then when we can't even get a thank you. I am not saying that this is all of our patients or family members, but there will always be some. Sometimes they take so much and we receive so little in return. Then you must learn to appreciate yourself and the things you have accomplished, and when you do get appreciation, remember those moments. Remember you are making a difference.

One of the lowest lows as a nurse is the feeling of isolation. The feeling of being in a constant state of loneliness. Nursing is a hard profession and can be the most isolating. Not being able to share everything that happens to us during the day, and the belief that we cannot share the things we have to face because we don't want others to have to feel the way we feel, or we don't want to feel like a burden to others. As nurses, we become so focused on taking care of others that we tend not to realize how lonely we are until it's too late. We have to learn to spot the feelings of isolation and understand that we are not alone in what we face and that through all the lows and the challenges we face, the thing that keeps us going is the hope of making a difference, even if it is just a small one.

Every day, we wake up knowing we have chosen a career where we get to help people for a living. How amazing is that? We get to be there for people, sometimes when no one else is. We have the opportunity to literally save a life. How many people get to say that? The joy we can bring to patients and families after we tell them good news. Or just doing the little things for them, such as taking them for a walk, helping them shower, or even just listening to what they are going through and the feeling we get inside when what we do actually makes a difference.

A fellow nurse shared a touching story about a time when she made a profound difference for a patient simply by being present and offering comfort. The patient needed to be transferred and was waiting for a flight—a process that can be lengthy in critical access facilities. During the wait, the patient became overwhelmed with worries about the financial burden and fear of flying, as it was her first time. The nurse's role shifted from medical care to emotional support. By listening attentively, acknowledging the patient's concerns, and talking her through her anxieties, the nurse provided the crucial reassurance needed.

As nurses, much of our work involves physical care, but equally important is our role as a support system for patients—a lifeline they can rely on regardless of the issue. It's remarkable how strangers in these moments can feel like family. We often find common ground with patients and have the privilege to help them, even in seemingly small ways that make a world of difference to them.

We expand our knowledge not just in nursing but in life. We get to meet people from all different backgrounds, or cultures, and get to create connections with the people we care for. As a nurse, we get to have the opportunity to take care of people from all over the world and get to experience their cultures with them and learn more than just medicine; we get to learn about life. We get to help our patients both in the medical aspect and also get the opportunity to take care of their cultural beliefs as well. We as nurses get the opportunity to not only support their physical health, but also their mental and emotional health. We get to heal people not only from the outside in, but also from the inside out.

As nurses, we actually get to use our knowledge learned, to help care for our patients. I'm sure we can remember going through nursing school and all the books and hours spent in the classroom trying to stuff as much knowledge in as we could. Although at the time you probably felt like you would never remember the things you learned, you will. When you get to actually practice the skills you learned, you start remembering things and are able to actually use your knowledge. That feeling you get when a patient comes in and you know what's going on and you are able to start anticipating the treatment, it's like a rush. You finally feel as though all those hours spent studying and all those tears you cried are all worth it. You did it! You're a nurse. You're making a difference.

Along with getting to use the knowledge learned in school, we get to learn something new every day, as long as we are willing to allow ourselves to. In our field, there is always knowledge to be learned, whether that's in our specific area or in another. It's a never-ending cycle of knowledge, and as long as we set our egos aside and are always willing to relearn something or learn something for the first time, we will always succeed. Not just in nursing but in life. Knowledge is power. The more we can know and understand, the bigger the difference we can make.

As nurses, we have the privilege of making a positive impact every day, whether through helping, empowering, or leading others. Beyond our role in healing patients, we possess the ability to empower and educate them. It's our responsibility to ensure patients understand their rights, grasp their medical conditions, and advocate for their own care. By giving them choices and encouraging them to speak up, we empower them not only in their healthcare decisions but also in their daily lives.

For many patients, our guidance may be their sole voice in navigating their daily challenges. Sometimes, we may not realize the profound effect we have on them, but our actions make a significant difference. Nurses are not only caregivers but also leaders and role models within their communities. People look to us for answers and reassurance, presenting us with opportunities to create meaningful change, not just for those in our care but for society as a whole.

We get to share the amazing experience of helping people with coworkers who start off as strangers but eventually become family. All the craziness and weird things that happen to us turn into laughs and inside jokes. From sharing laughter because you got pooped or peed on, making inappropriate jokes about things that have happened, or just sharing our craziest stories with each other. Although nursing is one of the most draining careers, it's moments like those with your peers that get you through them. Sometimes laughter really is the best medicine.

For some nurses, their high of nursing may be getting to save and help people now when before they couldn't. As discussed in my introduction, every single nurse has some reason they got into nursing. For some, it's so they can be there for people now, so others might not have to go through the same thing they went through, or so they can help patients feel the way another nurse might have made them feel. Such as in my path to nursing, I chose this career so I could help patients the way those nurses helped my aunt, giving people a sense of life even when there was no hope.

We have all experienced different paths in nursing and have all faced different ups and downs, but when faced with the lows and all the twists and turns, we must remember the highs and all the good things that have come out of this career. Remembering all the things that we are able to accomplish and all the people we have impacted. Understanding that, in the end, the good does outweigh the bad, and eventually this roller-coaster ride will even out, and all you will remember is the fun you had and the people's lives you changed.

This next page is a blank page. Use a black pen. Write down all the things that bother you, and let it disappear; not only on paper, but in your mind. It's time to let that shit go.

What You Do Matters.
You Matter.

The difference you have made

Chapter 7
Finding the Passion

Whether you're a nursing student or a nurse, I am sure we all have questioned once or twice why we have decided to become a nurse. The truth is, no one can answer this question besides you. We have all come into this field for different reasons, whether it is personal, if we like helping people, it's a stable career, or all of the above. All of us have different reasons and are all at different stages in life, so it's up to us to find the passion that keeps us going.

Nursing is and will continue to be one of the hardest and most draining careers, but also the most rewarding. Nursing is the type of field where it doesn't matter what your age, background, or family is. You always have a chance to succeed. There are so many different areas or units to have success in. You can be a clinic nurse, in-home nurse, hospice nurse, hospital nurse, specialty and so many more. As long as you're willing to put in the work, the opportunities are endless.

It's hard work, whether you're in nursing school or in the field; it's hard and not every day is easy. Truthfully, most of the time, you have more hard days than good days. That's when the passion needs to come in. You need to find out why you want to do this job. Having to deal with patients on their hardest days every day is not easy and can be draining, but if you have a true passion for it and a true reason for this job, you will succeed. A nurse told me that she keeps going because she gets to be her patients' rock. She gets to be her patients' advocate and gets to protect their wishes, which allows her to feel like she is making a difference. No matter how many years of nursing, we all need something, to get us to keep pushing forward, to help us feel like we are making a difference.

Reaching out to coworkers so they can become friends is also important. If you have a good working relationship with your coworkers, you will be happier to go to work every day. During some shifts, the only way you can get through them is to laugh about them with the people you work with. We are a team. We suffer together and we laugh together, and in some cases, we have to be each other's support. A good atmosphere helps create more happiness and the more happiness you have at your job, the more passion you will have, to continue.

Besides having a good support system around, we have to learn to improve our confidence. The best way to build confidence is to set one large goal; with small goals that can be achieved daily to get to that goal. Each day you succeed at something small, your confidence grows, until one day you get to that large goal and it doesn't seem that large anymore. In nursing, it can be hard to have confidence when some sort of defeat happens almost every day. By setting your daily goals small, they become easy to achieve and if you are able to have some sort of success amongst the defeat, you can continue to build confidence. When we regain confidence, we regain passion.

Learn to celebrate the small victories in your career. As a nurse, it's easy for us to get swept up in the negative and forget about the positives. We have to learn that small victories are still wins and they deserve to be celebrated. For example, getting to put your first NG tube in might not seem like such a rewarding task, but remember, you got trained to do that skill, and you just succeeded at it for the first time. How cool is that! Passion surrounds us in this field. We just have to learn what it looks like. Remember, sometimes it just takes something small to open a whole new world of discovery.

Sometimes we have to rediscover our passion. We have to allow ourselves to relearn the 'why' behind what we do. For some, it could be just taking a break and practicing our self-care, or taking a new grad under your wing and becoming a mentor. For others, it might be changing the area we work in, to allow ourselves to try something new. Rediscovery does not look like one thing. It is many different roads and it doesn't matter what road you take, as long as you get to where you're going.

For some, you just might not love nursing anymore, but you can't imagine doing anything else. That's okay. Not everyone loves their job, but you can still find love in how you can help people. What we do matters, but it's up to us to discover why we do what we do, in order to continue doing what we do. Nursing is a challenging career with a lot of ups and downs, but in the end, it's all worth it. It's worth all the pain and suffering that can come with it, because the patients we're able to help and heal, allow us to heal ourselves, and knowing that, even if we can just make a difference in one person's life, we are making a difference in the world.

"As a nurse, we have the opportunity to heal the heart, mind, soul, and body of our patients and ourselves. They may forget our name, but they will never forget how you made them feel."

– Maya Angelou.

Words of Wisdom

"You will never be prepared for the things you will see, no matter how many times you see them."

– Mckenna, 2 years

"Don't get too many irons in the fire. Make sure to mark out personal time, and don't let work infringe upon that personal time. You need that personal time to refuel yourself, otherwise, you will burn out. It is hard to take care of others if you are burned out."

– Anita, 37 years

"It is not your job to know everything, but it is your job to know what you don't know."

– Brittney, 2.5 years

"Nursing is like a roller-coaster. It has its ups and downs and round and rounds but in the end, it always smooths out."

– Mackenzie, 2.5 years

"Many of life's failures are people who did not realize how close they were to success when they gave up." – Thomas Edison.

"It really hits home because no matter how hard it is or if you're set back in school, you just need to push through because at the end of the day, it's worth it."

– Keala, nursing student

"Nursing school does not prepare you for what you will see and do. Learn and grow from your mistakes and don't be afraid to ask for help."

– Christine, 2 years

"The most important patient you look after in your career is yourself."

– George, 20 years

"Take care of yourself. You are replaceable to your employers, but not to your family and loved ones. It is okay to say no! It's okay to rest. It's okay to take time off."

– Deidre, 19 years

"Don't take things so seriously and don't be so hard on yourself."

– Denise, 16 years

"Find the grace to care for yourself as you have cared for others."

– Kristine, 17 years

"Nursing or being a nurse is not an easy job. To stay in this kind of work, you need to have a lot of patience and a whole world of compassion."

– Daisy, 17 years

"Make sure to build a good solid trustworthy relationship with your doctors, and don't be afraid to ask questions."

– Sarah, 13 years

"Reality is the best job you can do with the knowledge you have; that's all anybody can ask of you."

– Brenda, 44 years

"Knowledge is power. The more you know, the more you realize you don't know that much. Always be open to learning."

– Deana, 20 years

"One patient at a time, one day at a time."

– Eunice, 2 years

"Compassion isn't compassion if you leave yourself out."

– Stephanie, 2.5 years

"We have the profound privilege and responsibility of caring for women when they are simultaneously at their strongest and most vulnerable."

– Deanna, OB nurse, 29 years

"What we do today will write a memory in the heart of a family. That story will be told over and over – good or bad. What story will they tell about you?"

– Sheila, 30 years

"Imagine that is your daughter, friend, family you're caring for. No matter what their story is, they deserve the best, most respectful care always."

– Laura, 16 years L&D

"When you start nursing, keep a journal, so at the end of your shift if you learned something new, had a funny patient or a bad experience, you can write it down. Some days, if you can just write a few things down, you can get it out of your head so you can join life again."

– Betsy, 10 years

"Be teachable."

– Tamara, 23 years

"Remember your goal. Why are you here? Why did you start this?"

– Kristel, 2 years

"Relax. Breathe in and out. Don't let the anxiety hit you. Ask for help."

– Flor, nursing student

"Be willing to accept constructive criticism. It's not meant to bring you down. It's meant to shape you into a better nurse."

– Chantal, 2 years

"Every day is a learning process."

– Jenny, 27 years

"Don't be afraid. Follow your dreams; it's well worth it."

– Sweet, 13 years

"Celebrate your accomplishments."

– Chantal, 2 years

"If you think you know it all, you will kill someone."

– Theresa, 23 years

"Know enough to know you don't know much."

– Hannah, 3 years

"Don't be afraid to admit that a unit isn't right for you. That's the beauty of nursing; all the versatility. Also, don't be afraid to admit when it's time to move on."

– Kelsey, 14 years

"Be aware of the overconfident. Most often, they don't know as much as they want you to believe, and will drag you down."

– Linda, retired after 47 years

"There will be horrible days. Go back in the next day; it gets better."

– Teresa, 17 years

"It's called the practice of medicine because none of us are perfect you are human, and it's okay to cry."

– Kristi, 24 years

"Develop a routine for everything. If you follow the same steps every time, you will never worry about skipping something. It builds confidence and improves your quality of care."

– Leianne, 22 years

"Always take pride in the simplest of things, whether it's a bed bath or a simple smile. It all matters."

– Jade, 10 years

"Treat the patient, not the monitor."

– Raquel, 13 years

"Never stop caring. The moment you stop caring is the moment you should no longer be a nurse."

– Cari, 25 years

"With all the human suffering we bear witness to and the intimacy of what we are part of, it is easy to feel overwhelmed. Be gentle with yourselves. Please remember that we are all human. It's our compassion and empathy that called us to nursing. It's normal to have reactions and emotional responses to things. While we must stay focused and manage challenging situations and cases, we must also take time to accept and process those very human responses. Compartmentalizing or denying them can lead to becoming emotionally distant, anxious, depressed, and burnt-out. We are all in this together for the greater good. Stay calm in the chaos. Stay humble in the amazing moments saving lives. Be gentle with yourselves and your team when things are challenging or the outcome is difficult."

– Brenda, 24 years

"You only have two hands. You do your best and it will be what it will be."

– Heidi

"Stand up for yourself, stand up for your patient, stand up for yourself to the patient, listen to that voice and feeling that, 'something just feels off'. If the doctor won't listen, find someone who will."

– Casey, 9 years

"Always, always advocate for your patient, and remember, it really is our privilege to care for them."

– Dianne

You are courageous

You are outstanding

You are appreciated

You have so much more room for knowledge

You are full of more knowledge than you know

What you do matters

You are a nurse

References

Compassion Fatigue: Signs, Symptoms, and How to Cope. Canadian Medical Association. (2020, December 8). https://www.cma.ca/physician-wellness-hub/content/compassion-fatigue-signs-symptoms-and-how-cope

Hudson, J. (2023, May 1). *Achieving Work-life Balance in Nursing: Strategies for Success*. Better Nurse. https://betternurse.org/work-life-balance-nursing/

Importance of Self-care for Nurses and How to Put a Plan in Place. Purdue Global. (2019, February 13). https://www.purdueglobal.edu/blog/nursing/self-care-for-nurses/

Lee, K. A. & Friese, C. R. (2021, August). *Deaths by Suicide among Nurses: A Rapid Response Call*. Journal of Psychosocial Nursing and Mental Health Services. https://www.ncbi.nlm.nih.gov/pmc/articles/PMC8344804/

Mozafaripour, S. (2020, July 30). *Nurse Burnout: Risks, Causes and Precautions for Nurses*. University of St. Augustine for Health Sciences. https://www.usa.edu/blog/nurse-burnout/

Team, T. N. B. (2020, July 18). *Funny, Shocking and Silly Nursing stories – A Live Feed – The Nurse Break*. The Nurse Break. https://www.thenursebreak.org/breakroom-funny-stories/

Understanding the 6 Stages of Grief. Joshua York Legacy Foundation. (2023 February 14). https://www.joshuayorkfoundation.org/blog/the-6-stages-of-grief/

Vaughn, Natalie. (2023, July 17). *Make Time for Meaningful Recognition in Nursing*. Nurse.com Blog. https://www.nurse.com/blog/make-time-for-meaningful-recognition/

Washington, N., Martin, kim, Rangiris, A. L., & C, J. (2019, March 3). *My First Heartbreak as a Nurse*. Nurseslabs. https://nurseslabs.com/my-first-heartbreak-as-a-nurse/

World Health Organization. (2022, March 18). *Nursing and Midwifery*. World Health Organization. https://www.who.int/news-room/fact-sheets/detail/nursing-and-midwifery